THE AUDUBON SOCIETY POCKET GUIDES

A Chanticleer Press Edition

John Farrand, Jr.
Natural Science Editor
Chanticleer Press, Inc.

# FAMILIAR INSECTS AND SPIDERS

Alfred A. Knopf, New York

This is a Borzoi Book
Published by Alfred A. Knopf, Inc.

Prepared and produced by Chanticleer Press, Inc.,
New York.
Color reproductions by Nievergelt Repro AG, Zurich,
Switzerland.
Typeset by Dix Type Inc., Syracuse, New York.
Printed and bound by Dai Nippon, Tokyo, Japan.

First Printing.

Library of Congress Catalog Number: 87-46019
ISBN: 0-394-75792-0

Trademark "Audubon Society" used by publisher under
license from the National Audubon Society, Inc.

# Contents

**How to Use This Guide**   Insects and spiders are almost everywhere and are easy to see. Getting to know them will add interest to any walk outdoors and will increase your appreciation of other aspects of nature, since these tiny creatures have a great influence on the trees, flowers, and wildlife around them.

Coverage   This guide describes and illustrates 80 of the most abundant and colorful insects and spiders in North America, from Canada to Mexico and from the Atlantic coast to the Pacific. An additional 13 insects are mentioned as similar or related species.

Organization   This easy-to-use pocket guide is divided into three parts: introductory essays and drawings; illustrated accounts of the insects and spiders; and appendices.

Introduction   As a basic introduction, the essay "Identifying Insects and Spiders" tells you what characteristics to look for, and suggests questions to ask yourself when you see an unfamiliar insect or spider. "Watching Insects and Spiders" gives practical clues about where and when to go and what to do in order to see these often abundant and beautiful creatures. Black-and-white drawings show the important parts of insects and spiders.

The Insects and Spiders   This section includes 80 color plates arranged visually by shape, color, and overall appearance. Spiders and all the major groups of insects are represented. Usually a single species is illustrated and described; some large groups with very similar species are treated together either as a genus, family, or order. Facing each color plate is a description of the important field marks of the species or group, as well as information about its life cycle, habitat, and range. An introductory paragraph provides interesting facts about each species' habits, history, or close relatives. For quick reference, a drawing indicates the order to which the insect or spider belongs.

Appendices A special section "Guide to Orders" describes and illustrates the 17 major groups of insects and spiders covered in the guide. Knowing the distinguishing features of these broad categories will often help you to identify an insect or spider more quickly and easily. A glossary defines commonly used special terms.

Wherever you live there are colorful and interesting insects and spiders to see and enjoy. The more you know about them the more you will appreciate the world around you.

## Identifying Insects and Spiders

Most insects and spiders are relatively nearsighted, and if you move slowly and quietly, you should have no trouble getting a good look. A quiet approach offers two advantages: You can see features that will help you to identify the insect or spider, and you can have the fun and fascination of watching it go about its activities.

Basic Group

In identification, the first thing you should do is to determine an insect's or spider's basic group. Members of these basic groups, or orders, differ in a number of important ways, and even if you have never tried to identify an insect before, you may have no trouble deciding whether what you are looking at is a beetle, a grasshopper, or a butterfly.

Sometimes the order an insect or spider belongs to is not so obvious. There are harmless moths that look so much like wasps that predators stay away from them to avoid being stung. Unless you know the identifying features of moths, you too might be taken in. There are spiders that mimic ants so closely that you have almost certainly seen them and assumed they were ants. You should know what it is about the creature you have found that makes it a beetle, or a grasshopper, or a spider. This means taking a really close look.

8

# Parts of a Beetle

Thorax

Antenna

Head

Wing cover

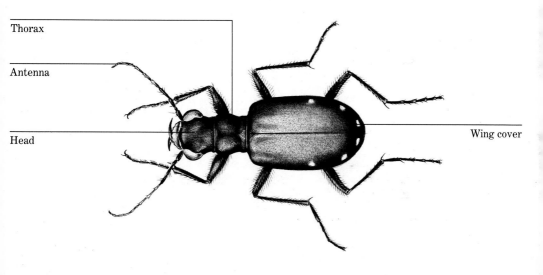

# Parts of a Grasshopper

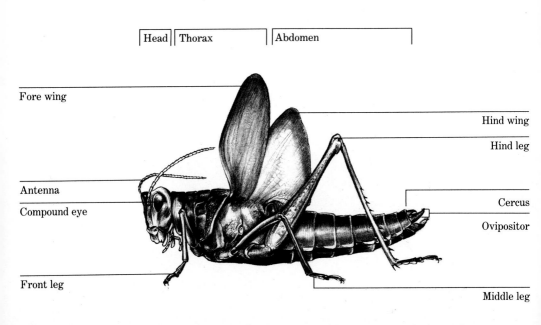

Head | Thorax | Abdomen

Fore wing

Hind wing

Hind leg

Antenna

Compound eye

Cercus

Ovipositor

Front leg

Middle leg

# Parts of a Butterfly

Fore wing

Fore wing

Antenna

Hind wing

Thorax

Mouthparts

Abdomen

# Parts of a Spider

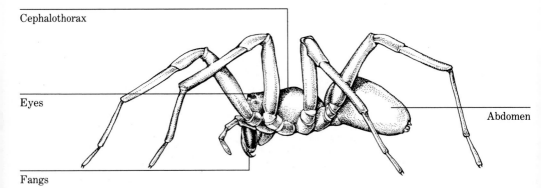

Cephalothorax

Eyes

Fangs

Abdomen

**Legs and Wings**  Ask yourself a number of questions as you study the insect or spider. How many legs does it have? Six legs immediately says you have seen an insect; eight legs means it is spider. Does the insect have four wings, two wings, or no wings? If it has wings, are they clear and membranous, opaque, or covered with scales? Do the wings have many veins or only a few?

**Head**  Next look at its head. Does the creature have antennae, or "feelers"? If it does, are they threadlike, feathery, or knobbed? What sort of mouthparts does it have? Are there two jaws, well suited for biting or chewing? Or are the mouthparts modified into a "beak" or a coiled tube? Are there no mouthparts, as far as you can see?

If you answer these questions, the odds are you will have gathered enough information to decide to which order the creature you have seen belongs. Then you can quickly check the text pages to find your insect or spider. For a more complete discussion of the orders of insects and spiders, see the "Guide to Orders" in the back of this book.

## Watching Insects and Spiders

From grasshoppers and crickets that sing to attract a mate, to bees, ants, and yellow jackets that live in populous colonies, to spiders that weave intricate webs to snare their victims, insects and spiders are fascinating to watch as they go about the often amazing details of their daily lives. Because these tiny creatures greatly outnumber all other animals and are abundant in all habitats, watching them requires little preparation, can take place practically anywhere, and offers hours of absorbing entertainment.

Habitat To see as many insects and spiders as possible, visit a number of different habitats. Gardens attract many species that feed on nectar or pollen at flowers. Leaf-eaters and their insect predators are also at home here. Meadows are the haunts of grasshoppers and crickets, wolf spiders that hunt in the dense tangles near the roots, and butterflies that flutter overhead. The margins of ponds and streams are good places to look for insects with aquatic young, such as dragonflies, damselflies, and mayflies, and for insects that inhabit the plants in these wet places.

Check dead trees for the Cottonwood Borer and other long-horned beetles, carpenter ants, and Pigeon

Horntails. Turning over a log will often reveal beetles, crickets, spiders, or an ant colony. Be sure to return the log gently to its original position after you have seen these creatures.

Time    You need not rise at dawn like a bird watcher in order to see insects and spiders. Insect activity reaches its peak at midday when the temperature is at its hottest. But a few insects, such as midges and mosquitoes, are most active at dawn or dusk. Still others, among them many moths and katydids, are busiest at night. Plan your activities accordingly.

Equipment    A ten-power magnifying glass or hand lens is the only piece of equipment you may need to watch insects. With it you can see all the details of the insects and spiders you find.

Watching insects and spiders is fun, easy, and rewarding. The variety of creatures you can find even in your own backyard rivals what you might see in the most exotic and remote wilderness areas and game parks.

# THE INSECTS AND SPIDERS

## Spur-throated Grasshoppers *Melanoplus*

These grasshoppers are the commonest ones in North American fields. In the 19th century, some western species formed immense swarms and devastated crops over wide areas, and even today, species of *Melanoplus* do more damage on agricultural land than any other grasshoppers. But on a sunny afternoon in midsummer, the soothing trill of one of these locusts, coming from somewhere in the warm, lush grass, is enough to make one forgive them their crimes, past and present.

Identification  ¾–1¾". Yellow, pale or dark brown, reddish, or green. Inner part of hind legs often banded or blotched; outer part often red or yellow. Antennae short.

Habitat  Grassy meadows, fields, and weedy places.

Range  Throughout the United States and S. Canada.

Life Cycle  Female inserts several egg masses in soil; nymphs mature in summer. Usually 1 generation a year.

### Carolina Locust *Dissosteira carolina*

Children in the country often call this big locust the "road duster," an appropriate name for an insect that is frequently found on dusty roads, is the concealing color of dust, and creates a little puff of dust when it plops back to the ground after one of its hovering flights. A liking for roadways has probably made the Carolina Locust much more abundant than it was before the forests were cleared by settlers.

Identification 1½–2″. A large grasshopper with short antennae. Tan or grayish, sometimes with a rusty tinge. Hind wings black with broad yellow margins, conspicuous in flight.

Habitat Roadsides, dry fields, vacant lots, and dunes.

Range Throughout North America.

Life Cycle Female places egg masses in soil. Eggs hatch in fall; nymphs overwinter and mature in following summer. One generation a year.

### Field Crickets *Gryllus*

Beginning early in spring, one or another kind of cricket can be heard calling from nearly any grassy place, and the cheerful chirping goes on all summer long. Even after the first frost, when a hush falls over the insect world, an occasional cricket can still be heard calling slowly in a field. When really cold weather sets in, a few enterprising crickets come indoors and continue their singing for a while, reminding us of the warm season just past.

Identification    ⅝–1". Black or dark brown and somewhat flattened, with dark antennae that are longer than the body and 2 long cerci projecting from the end of the abdomen.

Habitat    Meadows and weedy places.

Range    Throughout North America.

Life Cycle    One generation a year in North; eggs or nymphs overwinter in soil and adults appear in following spring or summer. Two or 3 generations in South.

### Snowy Tree Cricket *Oecanthus fultoni*

One of the most beautiful nighttime sounds is the rhythmic, late-summer calling, produced by rubbing the wings together, of male Snowy Tree Crickets. Often all the males in a thicket or hedgerow call in unison, creating a mellow, harmonic effect. The call of the Snowy Tree Cricket is slower when the air is cold, and can be used to measure the temperature. If you count the number of chirps in 15 seconds and add 40, you will have the temperature in degrees Fahrenheit.

Identification   ½–⅝″. Pale green and slender, with long antennae. Male has broad, transparent wings that cover the whole abdomen; female has narrower wings. Call a soft, rhythmic *treet, treet, treet, treet.*

Habitat   Woods, thickets, and garden shrubbery.

Range   United States and S. Canada; absent from southeastern states.

Life Cycle   Eggs overwinter under bark of twigs; nymphs mature in July or August. One generation a year.

24

### **Buffalo Treehoppers** *Stictocephala*

These little insects feed on plant juices by drilling through the bark with their "beaks." They are well named because they are often found in trees and can hop, and with their hump and two "horns," these treehoppers do look very much like miniature green buffalos. The adults prefer apple trees and elms. If they become numerous they can damage fruit trees, but most of them live in the woods and mind their own business.

Identification ⅜". Stocky and green or yellowish. Thorax with buffalolike hump prolonged over abdomen and 2 "horns" on each side. Eyes rather large and bulging. Transparent wings held along sides of body.

Habitat Woodlands, thickets, weedy meadows, and gardens.

Range Throughout the United States and S. Canada.

Life Cycle In late summer, females insert 2 eggs in C-shaped slits in twigs. Eggs hatch in spring; nymphs drop to ground, feed on clover, asters, and other plants, mature by midsummer, and return to trees to lay eggs. One generation a year.

### True Katydid *Pterophylla camellifolia*

Like most relatives of the grasshoppers and crickets, the True Katydid is best known for the call of the males— loud, rasping notes that begin to sound from broad-leaved trees in late July and continue until the first frost. At the height of the season the chorus can be deafening, but by fall few males are still calling, and in the cold night air they call more slowly and, it almost seems, dejectedly. The ones still calling are those that have failed to find a mate. Although this katydid is common, it tends to stay in the trees and is seldom seen.

Identification
: 1¾–2⅛". Bright green with fore wings oval, convex, and crossed by many heavy veins. Antennae long. Other large green katydids have the fore wings more pointed or longer and more oblong.

Habitat
: Forests, woodlands, and shade trees.

Range
: E. United States and S. Canada.

Life Cycle
: Eggs overwinter on bark of twigs and hatch in spring; nymphs mature in late summer. One generation a year.

**Praying Mantis** *Mantis religiosa*

A hunting mantis holds its front legs folded as if it were praying, but this large insect is a skilled predator, so some people think it should be called the "preying" mantis. There are several mantids in North America, but this introduced European species is one of the most common. It is easily recognized by a white spot under the base of each front leg. The Chinese Mantid (*Tenodera aridifolia*) is larger and lacks the spot under the front leg.

Identification 2–2½". Slender and green or tan. Front legs enlarged, armed with spines for grasping prey, and have a white, black-edged spot near the base; other legs long and thin. Abdomen long and covered by wings in adults. Head triangular with bulging eyes.

Habitat Meadows, fields, thickets, and gardens.

Range E. United States and S. Canada.

Life Cycle Female lays eggs in hard, papery mass in fall; nymphs hatch in spring, mature in late summer or early fall. One generation a year.

## Luna Moth *Actias luna*

Where there has been no extensive spraying, this lovely moth still pays quiet visits to window screens on still summer nights. But over much of its range, it has become hard to find. Even under the best conditions, the number of Luna Moths is small, and a single planeload of pesticide can wipe out the species for decades.

**Identification** 3⅓–4½″ wingspan. Wings pale green, outer margins purplish or yellow in some individuals. Hind wings with long tails; fore wings with purplish leading edge. Caterpillar green with yellow stripe on each side, red tubercles, and hairs.

**Habitat** Broad-leaved forests, especially those with hickory or walnut.

**Range** E. United States and S. Canada.

**Life Cycle** Adult emerges in April from cocoon in leaf litter and lays eggs on hickory, walnut, and a few other trees; caterpillar makes cocoon of silk and leaves, emerges as second generation of adult in July.

### European Cabbage Butterfly *Artogeia rapae*

This little immigrant from Europe first appeared in Quebec in 1910 and has since taken the country by storm, becoming the most abundant butterfly in many areas. Because of the caterpillar's taste for cabbages and other vegetables, it is a crop pest. But it also likes many weeds, and in cities is often the only butterfly you see, as it flutters along the sidewalk inspecting bits of greenery growing in cracks in the pavement.

Identification 1⅛–2″ wingspan. Off-white. Fore wings with blackish tips and 1 black spot at rear edge (2 in female); hind wings with pale yellow undersides. Caterpillar green with yellow stripes on back and sides.

Habitat Gardens, agricultural areas, vacant lots, and fields.

Range North America except Far North.

Life Cycle Spindle-shaped eggs are laid singly on cabbage and related plants—mustard, radish, watercress—and on nasturtiums. Up to 5 generations (April to November) a year; fewer in colder regions.

**Spring Azure** *Celastrina ladon*

Because this delicate butterfly usually flutters fairly close to the ground, in good habitat—where food plants for the caterpillars are abundant—you can see dozens in the air at one time. The Spring Azure is one of the first butterflies to appear in spring.

Identification ¾–1¼″ wingspan. Spring male has wings silvery blue above with border of small black checks; dark grayish below with small black spots. Spring female similar but with broad black border on fore wings. Summer generations paler above and below. This species never has orange spots on wings. Caterpillar variable, usually cream-colored; pupa golden brown.

Habitat Open woods, thickets, roadsides, and gardens.

Range Throughout North America.

Life Cycle Female lays green eggs on the flowers or flower buds of a variety of shrubs, including viburnum, blueberries, dogwood, and meadowsweet. Overwinters as pupa. Several generations each year in East; 1 or 2 in West.

### Large Wood Nymph *Cercyonis pegala*

These common butterflies vary widely over their extensive range. They all have two eyespots on the fore wing, but in some areas the fore wings lack the large yellow band. They fly erratically but with great skill, slipping quickly through thickets and weeds that lie in their path. Indeed, one can vanish before your very eyes just when you are chasing it to get a better look.

Identification 2–2⅞″ wingspan. General color light or dark brown with fine barring below. Fore wing near tip has 2 white or pale blue eyespots with dark borders, usually surrounded by broad band of yellow or buff visible from above or below. Caterpillar green with 4 yellow lines down back and 2 red "tails."

Habitat Open woods, thickets, weedy fields, and roadsides.

Range North America except Gulf Coast region.

Life Cycle Eggs laid on various grasses. Caterpillar hibernates soon after hatching, matures in following spring. One generation (June to September) a year.

### Silver-spotted Skipper *Epargyreus clarus*

The little yellow or orange skippers flutter peacefully through the grass, but this big, boldly marked species always seems nervous—dashing here and there or resting on a leaf for a moment before darting away again. If you are standing near a black locust tree and see a butterfly that races out in a frenzy and then races back into the foliage again, the odds are that it is a Silver-spotted Skipper.

Identification | 1¾–2⅜″ wingspan. General color dark brown. Has buff band across upperside of fore wing and a conspicuous silvery white spot on underside of hind wing. Full-grown caterpillar greenish yellow marked with bright green and with 2 red spots on dark brown head.

Habitat | Open woods, grassy hillsides, and gardens.

Range | Throughout the United States and S. Canada.

Life Cycle | Eggs laid singly on leaves of black locust, honey locust, and related plants. Caterpillar weaves shelter among leaves of food plant. One generation (May to September) in North; 2 or more generations in South.

### Red Admiral *Vanessa atalanta*

Male Red Admirals are noted for defending territories. From a lookout in a conspicuous place—often on a patch of sunlight under some trees or on a piece of white paper—they watch for any sign of movement. They dart not only at passing Red Admirals but at other butterflies, moths, and even people. Watch out for this when you walk through the woods; you may be next.

Identification    1¾–2¼″ wingspan. Fore wings dark brown with darker tips, broad orange band across tips, and several white spots. Hind wings brown with orange stripe along margins. Caterpillar spiny; color variable—black, brown, green—with yellow spots, white tubercles.

Habitat    Forest clearings, meadows, gardens, roadsides, and dunes.

Range    North America except Far North.

Life Cycle    Greenish eggs are laid on nettles and related plants; caterpillar feeds and pupates in folded leaf. Usually 2 generations a year, with adults from early spring to late fall; breeds year-round in S. California.

### Least Skipperling *Ancyloxypha numitor*

Like other small, yellow or orange, grass-haunting skippers, the Least Skipperling flies close to the ground and has a weak, fluttery flight. It is an eager flower visitor, and if you watch carefully when one of these diminutive butterflies lands on a flower, you will see an odd habit it has: Immediately after landing, it twirls its knobbed antennae a few times.

Identification ¾–1″ wingspan. Small size, uniform bright orange-yellow undersides of hind wings, the 2-toned (orange and dark brown) uppersides of all 4 wings identify this skipper. Wing tips broadly rounded. Caterpillar pale green with brown head.

Habitat Wet meadows, grassy marshes, and streamsides.

Range E. United States and S. Canada west to the Rockies.

Life Cycle Eggs laid singly on millet or bluegrass. Two or 3 generations (May to October) in North; 4 generations (April to November) in Texas.

44

## Tiger Swallowtail *Pterourus glaucus*

The commonest swallowtail in the East, this butterfly is usually seen in bounding flight near the foliage of trees or visiting flowers for nectar. West of the Rockies, the Western Tiger Swallowtail (*P. rutilus*) is similar, but the yellow spots in the black margin of the fore wings are fused into a yellow band; its caterpillar feeds mainly on willows, poplars, and aspens.

**Identification** 3⅛–5½″ wingspan. Yellow with black stripes and black borders spotted with yellow. Hind wings have a row of blue spots inside margin, a red spot at rear, and long black tails. Some females blackish but have same yellow, blue, and red spots as yellow females. Caterpillar green with orange-and-black eyespots behind head.

**Habitat** Forests, clearings, gardens, and roadsides.

**Range** Alaska and Canada southward east of Rockies to Gulf of Mexico.

**Life Cycle** Eggs laid singly on black cherry, yellow-poplar, and other trees; caterpillar feeds on foliage, then pupates in folded leaf. One to 3 generations a year.

46

**Monarch** *Danaus plexippus*

Many of the milkweeds that Monarch caterpillars feed on are poisonous, so the butterflies are poisonous, too. Their bold coloring serves as a warning to birds. The tasty Viceroy (*Basilarchia archippus*) mimics the Monarch and thus gains protection from predators. You can tell a Viceroy by its smaller size, thin black line across the hind wings, and habit of soaring with the wings held flat rather than in a shallow V like the Monarch.

Identification 3½–4″ wingspan. Easily identified by orange-brown color, black borders and veins on wings, and black body with white spots. Caterpillar striped black, white, and yellow with black filaments fore and aft.

Habitat Meadows, weedy fields, and roadsides, especially where milkweeds grow.

Range North America except Far North.

Life Cycle Eggs laid singly on milkweeds. Caterpillar matures in 10 days, then forms green-and-gold chrysalis. Adult emerges in 12 days. Several generations a year; last one migrates south for winter, flies north in spring.

48

## Apantesis Tiger Moths *Apantesis*

The 30 species of this group have special defenses against predators. Their bold patterns warn birds that they taste unpleasant, and keen hearing enables them to detect the high-pitched sonar of foraging bats in time to dart out of harm's way. Males often come to lights at night, but for some reason females are less often seen. Perhaps they fly mainly during the day or tend to stay close to the caterpillars' food plants.

Identification 1⅛–2″ wingspan. Furry, stout-bodied moths with fore wings white, yellow, or pink, boldly checkered black; hind wings pink, yellow, or whitish with black marks. Abdomen pink or yellow, often with black spots. Caterpillars hairy and black with pale line down back.

Habitat Meadows, fields, thickets, and roadsides.

Range Throughout North America.

Life Cycle Eggs are laid on various herbaceous plants; caterpillars overwinter in leaf litter, pupate in spring; adults fly in late spring or summer, or in spring and again in fall. One or 2 generations a year.

### Sheep Moth *Hemileuca eglanterina*

This handsome western moth likes open country and is easy to spot as it flies swiftly about during the day like a butterfly. It can be told from a butterfly by its stocky body and feathery antennae. The spines of the caterpillars can sting if you are not careful.

Identification 2½–2¾″ wingspan. Fore wings usually pinkish, marked with black spots and stripes; hind wings and stout, furry body yellow with similar markings. An all-black form is found on Mount Shasta in N. California. Caterpillar spiny; brown with a red stripe on each side and a row of red spots down back.

Habitat Meadows and pastures; only in mountain meadows in southwestern states.

Range W. United States east to Rockies.

Life Cycle Eggs are laid in clusters on twigs in summer; caterpillar feeds in groups on wild roses, coffeeberry, and other shrubs, then spends winter as pupa. Adult emerges in following summer or in summer after that. One generation a year.

52

## Underwing Moths *Catocala*

At rest on the mottled trunk of a tree, an underwing moth is beautifully concealed not only from people but also from birds looking for a meal. If you touch one of these moths by accident, it will quickly lift its dull-colored fore wings and flash the bold colors on the hind wings. The object is to startle you, or a bird, just long enough to allow the moth to make a hurried getaway, land on another tree trunk, and disappear again.

Identification  1⅖–3⅘″ wingspan. Fore wings mottled gray and brown; hind wings red and black, yellow and black, or in some species solid black. Caterpillars mottled gray, resembling twigs.

Habitat  Forests, woodlands, and gardens with large trees.

Range  Throughout North America.

Life Cycle  Eggs are laid in fall on a variety of trees, depending on the species, and hatch in spring; caterpillars feed on foliage, pupate in leaf litter, and usually emerge as adults in late summer. One generation a year.

54

## Hummingbird Clearwing *Hemaris thysbe*

Most sphinx moths pay silent visits to flowers at night, but the well-named Hummingbird Clearwing uses its long tongue to sip nectar in broad daylight. As it hovers in front of a flower on rapidly beating wings, this greenish species bears a definite resemblance to a hummingbird, which probably saves it from interference by predators. A closely related species, the Snowberry Clearwing (*H. diffinis*), avoids its enemies by looking like a large and menacing bumble bee.

Identification  1½–2″ wingspan. Adult has wings mainly clear with brown, scaled borders. Body furry with foreparts olive-green and most of abdomen dull rust-brown. Caterpillar pale green with darker green stripes, reddish-brown spots, and yellow horn on tail.

Habitat  Clearings, meadows, and flower gardens.

Range  S. Canada and N. United States; in East south to Gulf Coast.

Life Cycle  Caterpillar feeds on honeysuckle, viburnum, and other shrubs. Two generations a year.

## Woolly Bear Caterpillar *Pyrrharctia isabella*

One of the most familiar sights in the fall is a Woolly Bear Caterpillar hurrying over the ground, searching for a safe place to spend the winter. If you touch one, it will curl up like an armadillo, protected by its dense covering of stiff bristles. These caterpillars are hard to pick up because the bristles make them slide out from between your fingers.

Identification 2″. Covered with stiff bristles; black with a broad band of brown around middle. Adult moth has 2-inch wingspan; fore wings dull yellow with small black spots; hind wings paler with gray dots; 3 black dots on each segment of abdomen.

Habitat Roadsides, meadows, fields, and bare ground.

Range North America except Far North.

Life Cycle Caterpillar hibernates, then spins cocoon in spring. Adult emerges 2 weeks later and lays eggs in clusters on a wide variety of plants. These eggs hatch, caterpillars grow, pupate, and emerge as adults; offspring of this second generation hibernate as caterpillars.

**Eastern Tent Caterpillar** *Malacosoma americanum*

The communal web nests of this species are a familiar sight in cherry and apple trees in spring. Although there are a few every year, in some years these caterpillars reach plague proportions, and every black cherry loses all its leaves. In the West, the Pacific Tent Caterpillar (*M. constrictum*) has bold blue spots, and the California Tent Caterpillar (*M. californicum*) lacks the blue spots.

Identification 2″. Body with many black hairs, yellowish stripe down back, blue spots, and reddish and yellow stripes along sides. Adult moth (wingspan 1½–1¾″) dull brown with 2 pale lines on each fore wing.

Habitat Woodlands, thickets, roadsides, and gardens.

Range E. United States and S. Canada.

Life Cycle Egg mass overwinters on twigs of black cherry and related trees; caterpillars hatch in spring, build "tent," and feed on foliage, then leave host tree to pupate in silken cocoon. Adults appear in early summer. One generation a year.

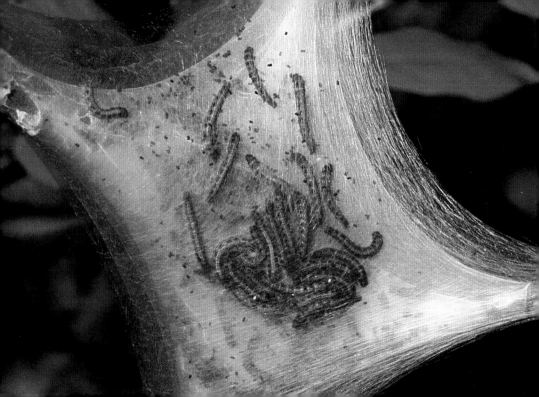

**Aphids** *Family Aphididae*

Soft-bodied and docile, aphids use their "beaks" to feed on the sap of plants. They are abundant, breed rapidly, and are often defended by ants that collect their sugary secretions. Aphids are eaten by green lacewings, ladybird beetles, the larvae of some hover flies, and tiny parasitic wasps. Without these predators, the world might well be overrun with aphids.

Identification 1/16–1/3″. Pear-shaped, small-headed insects with long antennae and often with 2 peglike cerci at end of abdomen. May be green, black, red, or other colors. Almost always found in dense clusters on plants.

Habitat Widespread on vegetation of all kinds.

Range Throughout North America.

Life Cycle Variable. Eggs survive winter, hatch into wingless females that give birth to more wingless females but no males, then to a winged generation that moves to another host plant; more wingless females, then migration back to the first host plant. Finally, males are produced, mating occurs, and eggs are laid.

62

**Cicadas** *Family Cicadidae*

A familiar summer sound is the buzzing or whining "song," produced by vibrating membranes at the base of the abdomen, of male cicadas. This eastern Dogday Harvestfly (*Tibicen canicularis*) sings during the hot "dog days" of August. Nymphs of the periodical cicadas (*Magicicada*) spend 17 years underground in the North, 13 in the South.

| | |
|---|---|
| Identification | 1–2½". Like a stocky, giant fly, but has 4 wings instead of 2, bulging eyes, large wings with strong veins often tinted green, yellow, or orange, and sturdy legs. |
| Habitat | Forests, woodlands, shade trees, and deserts. |
| Range | Throughout the United States and S. Canada. |
| Life Cycle | Females lay eggs in slits in bark of twigs; eggs hatch, nymphs drop to ground and feed on sap drawn from roots; nymphs remain underground for 1-17 years depending on species, then dig tunnel to surface in summer and molt into adult while clinging to a tree trunk or fence post. |

**Caddisflies** *Order Trichoptera*

Not much to look at, these small brown insects visit porch lights so frequently that most people have seen them. They are usually mistaken for moths, but no moth has such long, slender antennae. The cases built by the aquatic larvae are so distinctive that an expert can identify the species just by looking at the case.

Identification  $\frac{1}{32}$–1″. Slender, usually brown insects with 4 wings held rooflike along abdomen; long, slender legs. Resemble moths but have long, threadlike antennae and hairy wings. Females of a few species lack wings.

Habitat  Near lakes, ponds, marshes, and streams.

Range  Throughout North America.

Life Cycle  Eggs are dropped into water or attached to leaves over water; larvae build cases of sand grains or plant fragments, crawl over stream bottoms, or weave small nets to catch prey in current; larvae that build cases pupate in them. Adults are short-lived and rarely eat.

### Green Lacewings *Chrysopa*

These delicate green insects are most often seen around lights at night, where their feeble, almost struggling flight gives little hint that they are voracious predators that specialize in aphids. The squat larvae have the same appetite and can make a difference in the number of aphids that infest garden flowers and vegetables. If accidentally crushed, the adults give off a pungent odor—perhaps a defense against birds and other predators.

Identification    ⅜–¾". Pale green or yellow with glittering golden eyes. Wings long, clear, finely veined with green. Antennae long and threadlike. Larvae flattened and bristly with curved, sharp jaws.

Habitat    Woodlands, thickets, meadows, and gardens.

Range    Throughout North America.

Life Cycle    Eggs are laid in clusters on foliage, each one on a slender stalk. Larvae feed mainly on aphids, then pupate in silken cocoons attached to a leaf. One generation a year, with each species on a different schedule.

**White Tail** *Libellula lydia*

Male White Tails spend most of their time darting about and threatening one another by displaying their white abdomens, or lying in wait ready to charge out after a passing rival. Each male is determined to keep others out of its territory. This is a serious business, but watching a group of these dragonflies chasing and cavorting over a pond, one can easily believe that they are enjoying every minute of it.

Identification 1¾″. Male has gleaming white abdomen and a large black band across each wing. Female has brown abdomen with a row of yellow spots; a dark patch in middle of each wing and a larger dark patch at tip. Both sexes have flatter abdomens than most other dragonflies.

Habitat Marshes and weedy ponds.

Range Throughout the United States and S. Canada.

Life Cycle Female lays 25–50 eggs at a time by dabbing surface of water with tip of abdomen, often near a log or board, while male hovers nearby. One generation a year.

### Green Darner *Anax junius*

A male Green Darner patrols its territory—often a whole pond—by zooming back and forth through the air 10 or 15 feet above the water. It tolerates smaller dragonflies that fly closer to the surface but quickly drives away any other male Green Darner that enters its air space. When a female arrives, there is a brief skirmish, then mating takes place. Guarded by the hovering male, the female immediately lays her eggs.

Identification 2¾–3″. Easily identified by large size, green thorax, and blue or purplish abdomen. At close range shows targetlike mark on face above eyes. Wings clear, tinged with yellow near tips and darkening in older insects.

Habitat Ponds, open marshes, and sluggish streams.

Range Throughout North America; more common in East.

Life Cycle Female inserts eggs into tissues of aquatic plants. Naiads may spend more than one year in water, then emerge as adults in late spring or early summer.

## Damselflies *Suborder Zygoptera*

Less flamboyant than the larger, sturdier dragonflies, damselflies live closer to the surface of the water and fly more quietly through the vegetation. When at rest, they never hold their wings stiffly out to the sides as dragonflies do. But they too are skilled predators, capturing small flying insects. Some species are called bluets because of their color, others are known as dancers because of their agile flight.

| | |
|---|---|
| Identification | 1–2″. Slender and delicate with long, slim abdomen and small, broad head with bulging eyes. Thorax and abdomen often marked with bright blue or green. Some species red or glossy blue-black. At rest, narrow wings held together over back or fanned slightly. Naiads slender with 3 leaflike gills at tip of abdomen. |
| Habitat | Ponds, marshes, and shallow streams. |
| Range | Throughout North America. |
| Life Cycle | Eggs are inserted into tissue of aquatic plants; predatory naiads overwinter and emerge as adults in following year. |

## Mayflies *Order Ephemeroptera*

Few of us ever see the aquatic naiads of mayflies, but adults can be conspicuous when they swarm at mating time. Some mayflies appear in such numbers that they pile up under streetlights and on bridges and interfere with traffic. Adults live only a day or so and cannot eat. They are a favorite food of trout, and many dry flies tied by anglers are imitations of these delicate and short-lived insects.

Identification ⅛–1⅛″. Slender, soft-bodied, brown or yellowish insects with many-veined, triangular wings held like sails over back; 2 or 3 long, hairlike tails at tip of abdomen. Naiad flat with a row of leaflike gills along each side of the abdomen and 3 hairlike tails at tip.

Habitat Near rivers, streams, ponds, and lakes.

Range Throughout North America.

Life Cycle Naiads live 1–4 years in water, feeding on debris and small animals; last stage molts into winged form which soon molts again into adult. Mating and egg-laying quickly occur over water, and adults die.

### Crane Flies *Tipula*

These large, gangling flies often come to lights at night and are sometimes misidentified as giant mosquitoes. But they are harmless; the adults of most species can only lap up water and nectar from flowers. Their long legs break off readily, making it easy for us to identify them and helping them break free from spider webs, which are common in the damp, shady places they inhabit. The soil-dwelling larvae are sometimes called leatherjackets on account of their tough skin.

Identification   ⅜–2½″. Slender with small head and thorax, long abdomen, and long, slender legs, easily broken from body. Thorax has V-shaped groove on back. Wings long, held out to sides, often patterned.

Habitat   Moist woodlands and fields, gardens, and streamsides.

Range   Throughout North America.

Life Cycle   Eggs are laid in moist earth; larvae feed on decaying plant matter, usually overwinter as pupae. Adults emerge in spring. One to several generations a year.

**Stoneflies** *Order Plecoptera*

As their name implies, adult stoneflies are often found crawling on stones along the edges of streams. They have two pairs of wings, but they are weak fliers and often try to escape danger by scrambling away on foot. Adults are short-lived and usually eat nothing; the aquatic naiads are either predators or eat algae.

Identification  ¼–2½″. Adults long and slender, often blackish, with long antennae, wings folded flat over abdomen, and 2 long cerci at hind end. Naiads also have 2 long appendages at tip of abdomen, and feathery gills at bases of legs.

Habitat  Streamsides and shores of lakes and ponds.

Range  Throughout North America.

Life Cycle  Eggs are deposited in masses in water; naiads mature in 1–4 years, depending on species and water temperature; adults live 2–3 weeks.

## Scorpionflies *Panorpa*

You can easily recognize a male scorpionfly by its abdomen, which is curved upward and ends in a pair of bulbous claspers used during mating. This makes the insect look something like a scorpion, but it can't sting. Adult scorpionflies are harmless, rather inactive creatures, most often found standing around on leaves in damp, shady places. Their flight is weak. They feed on decaying fruit, pollen, and insects, both alive and dead.

Identification  ½–¾″. Adults brown with face prolonged into distinctive snout; antennae long and threadlike; 4 wings spotted or banded, with many veins. Tip of abdomen of male curls upward like tail of scorpion. Larvae caterpillarlike, spiny, with short, thick antennae.

Habitat  Moist woodlands, clearings, and gardens.

Range  E. and central North America.

Life Cycle  Eggs are laid in soil; larvae feed on insects and other organic matter, overwinter, then pupate in burrows. Adults emerge in summer. One generation a year.

## Midges *Family Chironomidae*

Although in England the word "gnat" usually means a mosquito, John Keats was probably referring to a mating swarm of midges when he wrote:

Then in a wailful choir the small gnats mourn
Among the river sallows, borne aloft
Or sinking as the light wind lives or dies.

These swarms produce a high-pitched whine from thousands of tiny beating wings and consist mostly of males.

| | |
|---|---|
| Identification | 1/16–3/8″. Adults mosquitolike but without long "beak"; head bent under humped thorax; antennae of males very feathery. Larvae of some species bright red. |
| Habitat | Marshes and ponds; woodlands, meadows, gardens near water. |
| Range | Throughout North America |
| Life Cycle | Eggs are laid in water after mating swarm; larvae eat decaying plant matter on bottom. Mature larvae float to surface, pupate; adults released at surface. |

84

### Mosquitoes *Family Culicidae*

Female mosquitoes attract our attention mainly because they need a lot of protein in order to lay eggs. They obtain this protein from the blood of reptiles, birds, or mammals, transmitting malaria, encephalitis, and other diseases in the process. But most species are just a nuisance and can be kept at bay with a dab of insect repellent. Males feed on plant juices and cause no trouble at all.

Identification  ⅛–½". Slender flies with a long, thin "beak" and narrow wings with hairlike scales along veins and margins. Antennae of males feathery. Aquatic larvae slender, usually with short breathing tube near hind end.

Habitat  Widespread, breeding in quiet or stagnant water.

Range  Throughout North America.

Life Cycle  Eggs are laid singly or in small "rafts" on water or on damp ground that will be flooded; wriggling larvae feed on tiny aquatic plants and animals, then molt into active pupae. Adults emerge at surface from floating pupae. One or more generations a year.

86

### Robber Flies *Family Asilidae*

Robber flies are fierce predators that chase down insects in flight or pounce on them at rest. They often perch on twigs, waiting for a victim to pass by. Few insects avoid their clutches; they have even been known to capture dragonflies, and one entomologist saw a small robber fly pounce on another insect, only to be seized in turn by a larger robber fly. The spines on their stout legs form a basket to hold their prey in flight, while they pierce it with their sharp, beaklike mouthparts.

Identification  ¼–1⅛". Spiny-legged flies, usually hairy, with a "beard" of bristles on face and a depression on forehead between eyes. Most are slender with tapering abdomen, but some are stockier and built like bumble bees.

Habitat  Fields, gardens, and woodlands.

Range  Throughout North America.

Life Cycle  Eggs are laid in soil, dry leaves, or decaying wood; larvae prey on other insects. Usually 1 generation a year.

### Tarantula Hawks *Pepsis*

The well-named tarantula hawks belong to a group of wasps that specialize in paralyzing spiders to feed their larvae. They are active and nervously flick their wings as if aware of the danger of taking on a hairy tarantula many times their own size. But spiders are easier to subdue than other prey because they have a major nerve center on the underside of the body; a single well-aimed sting can put them out of commission. The battle between a tarantula and one of these wasps can last a long time, but it generally ends in victory for the smaller and more agile wasp.

Identification ½–1⅘". Large, black wasps, often with reddish or orange wings. Females have curled antennae.

Habitat Deserts and dry grassy places.

Range California and southwestern states.

Life Cycle Females hunt for tarantulas, sting and paralyze them, dig burrows, and bury spiders and eggs. Larvae feed on spiders and pupate in burrows, then emerge as adults. One generation a year.

## Velvet-ants *Family Mutillidae*

These insects look like ants with a velvety or wispy coating of hairs, but they are really wasps. The wingless females, which are seen more often than the flying males, trot over the ground in search of the nest of a bee or wasp in which to lay their own eggs. It is best to let them keep trotting because they have a surprisingly potent sting that has earned one eastern species the somewhat exaggerated name Cow Killer.

Identification  ¼–1″. Densely hairy and boldly patterned in red, yellow, or orange. Antennae straight. Males have wings; females are wingless and look like brightly colored ants. True ants have "elbowed" antennae and a more obvious constriction between thorax and abdomen.

Habitat  Deserts and bare, dry ground.

Range  Throughout the United States.

Life Cycle  Females lay eggs in nests of bees or solitary wasps; larvae feed on larva of host, then pupate. One generation a year.

## Termites *Order Isoptera*

Unlike the social wasps, bees, and ants, termites have colonies containing both sexes. The queen's fertile consort is almost as pampered as she is. Often called white ants, they are abundant in warmer parts of the country. They feed on wood, digesting it with the aid of special microbes that live in their intestines.

Identification ¼–¾″. Small, soft-bodied, whitish or brown social insects with well-developed jaws. Resemble ants but abdomen broadly joined to thorax and antennae not "elbowed." Fertile, winged individuals usually blackish.

Habitat Underground or in dead timber.

Range Throughout the United States and S. Canada.

Life Cycle After brief mating flight, fertile male and female shed wings and establish colony; eggs and nymphs tended and protected by sterile worker and soldier castes of both sexes. Winged, fertile males and females produced at least once a year. When founding pair dies, they are replaced by fertile offspring. Colony may last for years, with millions of workers.

### Carpenter Ants *Camponotus*

Ants are among the most abundant and familiar insects. Carpenter ants are easily recognized when they are tunneling in logs or old buildings and producing piles of sawdust. Unlike termites, carpenter ants do not eat wood, but merely excavate it. Colonies can last for several years and grow very large, sometimes causing much damage to buildings. These hefty ants have a strong bite, but they cannot sting.

Identification Workers ¼–½″. Large, robust, blackish or reddish ants with stalk of abdomen one-segmented. Best identified by habit of nesting in galleries in wood.

Habitat Standing dead timber, buildings, and utility poles. Often forage in trees or on ground near nest.

Range Throughout the United States and S. Canada.

Life Cycle New queen establishes colony in dead wood and tends first brood of young; this first generation expands galleries and tends all future larvae and pupae; colony grows to thousands, including fertile young females and males, which leave galleries and perform mating flight.

96

### Black-and-yellow Mud Dauber *Sceliphron caementarium*

The Black-and-yellow Mud Dauber is solitary and does not nest in colonies. Like most solitary wasps, it is a gentle creature that seldom stings. Females are most often seen gathering mud for their nests at puddles, and both sexes feed on nectar at flowers. The Blue Mud Dauber (*Chalybion californicum*), entirely dark steel-blue, takes over the nests of this species, throws out the lawful occupants, and installs its own eggs along with spiders it has captured.

Identification  1–1⅛″. Glossy black with yellow legs and markings on thorax. Abdomen with slender stalk, either yellow or black. Wings blackish.

Habitat  Meadows with flowers, buildings, and rocky cliffs.

Range  Throughout North America.

Life Cycle  In spring, female builds mud nest with 2–6 cells under rocks or eaves; each cell contains 1 egg with paralyzed spider; larva feeds on spider and grows to ½″, then pupates in cell in brown, papery cocoon. Adult emerges in summer, mates, and females overwinter.

### Giant Ichneumons *Megarhyssa*

A parasitic female giant ichneumon uses its sensitive antennae to detect vibrations produced by a horntail larva deep inside a log. Then, looking something like an oil rig, it bores into the wood to lay an egg in the tunnel of its host. These are North America's largest ichneumons, a group containing more than 3,000 species, all of them parasites of a variety of other insects; the smallest ichneumons are only ⅛" long.

Identification   Female 1⅜–3". Body brown or blackish, banded with yellow or red and yellow. Legs yellow, often with dark markings. Females have long ovipositor trailing behind abdomen; males smaller, lack ovipositor.

Habitat   In forests, near dead timber.

Range   Throughout North America, except in treeless regions.

Life Cycle   Females drill holes in log or tree trunk, lay eggs in larval tunnels of Pigeon Horntail or related species; larva of ichneumon feeds on horntail larva, then pupates in tunnel; emerges as adult in following summer. One generation a year.

### Pigeon Horntail *Tremex columba*

In the fall, when an exhausted female Pigeon Horntail has deposited her last egg in a dead tree, she dies without even pulling her ovipositor out of the wood. Indeed, most of the Pigeon Horntails one sees are lifeless females clinging to tree trunks. The larvae have deep tunnels, and you might expect them to be safe from harm, but even here they are attacked by giant ichneumons.

**Identification**  1–1½". Large and wasplike but without slender "waist" between abdomen and thorax. Black or brown with dusky wings, yellow bands on abdomen. Female has spikelike ovipositor at tip of abdomen.

**Habitat**  Deciduous and mixed forests, especially on dead trees and logs.

**Range**  E. North America.

**Life Cycle**  Female pierces wood with sharp ovipositor and lays egg, along with spores of fungus; as larva grows, fungus softens wood; larva feeds on infected wood, pupates in cell under bark after 2 years. Adult appears in fall.

102

## Paper Wasps *Polistes*

The nest of these wasps is an exposed, papery comb of cells. Paper wasps will sting if disturbed at their nests, but they have a milder disposition than yellow jackets. A female that has established a colony with her own daughters as workers may be joined by females whose colonies have failed, but she makes sure that all the young raised in the nest are hers by eating any eggs laid by her helpers.

Identification   ½–1″. More slender than hornets or yellow jackets; reddish brown or blackish with pale bands and dusky wings.

Habitat   Meadows and fields with flowers. Usually nests on buildings.

Range   Throughout North America.

Life Cycle   Females start colonies in spring and rear unmated females that tend larvae in open nest of papery cells. In late summer, young males and females leave nest and mate. Males and older females die; young fertile females survive winter and start new colonies in following year.

**Hover Flies** *Family Syrphidae*

Few people dislike the colorful and inoffensive hover flies. The adults of most species visit flowers, where their resemblance to bees helps protect even the smallest ones from predators. Hover flies feed on nectar, a diet rich in the energy-giving carbohydrates they need to perform the spectacular feats of hovering that have given them their name.

Identification  ¼–¾". Variable. Often boldly patterned in yellow and black, or stocky, dark, and furry like a bumble bee. Differ from other flies in having a "false vein" (a fold that looks like a vein, except under a magnifying glass) in each wing. Usually seen hovering near flowers.

Habitat  Open fields, meadows, gardens, and woodland clearings.

Range  Throughout North America.

Life Cycle  Females lay eggs in a variety of habitats, including water, moist soil, clusters of aphids, and nests of ants and bees; larvae feed on other insects or decaying matter, and a few attack plant roots. Usually one generation a year.

### Yellow Jackets *Vespula*

A yellow jacket's bold pattern makes it easy to recognize. Experiments show that birds instinctively avoid insects banded with black and yellow or white. This protects not only the yellow jackets, which defend themselves and their nests with a sting, but also many unarmed but banded insects such as hover flies. If birds had to learn this the hard way, many yellow jackets would be killed before birds got the message.

Identification   Worker ½–⅝". Short, stocky wasps, boldly marked with black and yellow or (in a few species) black and white. Abdomen banded. Wings dark. Queens, most often seen in early spring, larger than workers.

Habitat   Woodlands, meadows, fields, gardens, and parks.

Range   Throughout North America.

Life Cycle   In spring, new queen establishes colony in underground burrow or in globular, hanging nest of paper, with sterile females as workers; fertile males and females appear in mid- or late summer, mating occurs, then males, workers, and old queens die. Mated queens overwinter.

### Honey Bee *Apis mellifera*

We humans rely on a dangerously small number of food plants for our survival. Just as important to us is the industrious Honey Bee, the major insect pollinator of our crop plants for more than 4,000 years. If anything were to happen to this one species of bee, hundreds of vital fruits and vegetables would quickly disappear.

**Identification**
Worker ⅜–⅝″. A medium-sized bee with thorax hairy and brownish, abdomen banded dull orange and black. Worker often carries yellow pollen mass on hind legs. Queen rarely seen; larger than worker.

**Habitat**
Croplands, orchards, gardens, fields, and woodlands.

**Range**
Throughout North America except Far North.

**Life Cycle**
Queen and sterile female workers overwinter in hive; in spring eggs are laid, workers feed larvae and tend pupae as number of workers grows as high as 50,000; in late spring or summer, fertile males (drones) and females appear, and old queen leaves hive with swarm of workers to start new colony, leaving daughter queen in old hive. Queens live 2–5 years, summer workers about 6 weeks.

### Bumble Bees *Bombus*

Even though they can sting, it is hard not to like bumble bees. They are peaceful and efficient gatherers of nectar and pollen, much too busy to bother anybody unless their nests are disturbed. Most of the bumble bees we see are workers, but in early spring and late summer you can find larger ones at flowers; these are fertile queens, the only ones that survive the winter to start new colonies.

Identification  ⅗–1". Large, furry bees, black with yellow or orange on thorax, part or most of abdomen, and sometimes head. Queens larger than workers.

Habitat  Woodlands, meadows, weedy fields, gardens, and bogs.

Range  Throughout North America.

Life Cycle  Hibernating queens appear in spring and establish underground colonies; workers—all sterile females—forage and tend larvae and pupae; late in season, males and young queens emerge, leave nest, and mate; males, workers, and old queens die; new queens overwinter and repeat cycle.

**Bee Flies** *Family Bombyliidae*

The sun-loving bee flies are experts at hovering and are most often seen at flowers sipping nectar through their long, beaklike mouthparts. If you go into the woods in early spring, before the leaves have cast their shadow on the forest floor, you may find a female bee fly hovering near a small hole in the ground, now and then making a pass at the entrance of the hole. She is probably tossing an egg into the nest of an insect that will serve as a host for her parasitic larva.

| | |
|---|---|
| Identification | ¼–⅝″. Stout, furry, beelike flies, usually with long "beak," and with wings (often patterned) held out at sides. |
| Habitat | Sunny meadows, fields, gardens, and open woodlands. |
| Range | Throughout North America. |
| Life Cycle | Females lay eggs singly or in near nests and burrows of other insects; larvae of bee flies feed on larvae of host, pupate in nest, and emerge as adults in following spring or summer. One generation a year. |

## Deer Flies *Chrysops*

When abundant, these flies can make being outdoors very unpleasant. Silent on the wing, a deer fly can land on its intended victim and begin to deliver its painful bite before it is noticed. As in the case of mosquitoes, only females bite; the males feed on nectar at flowers.

**Identification**  ⅜–⅝″. Flattened, blackish or dusty with yellow or greenish markings. Head small; eyes brilliant gold or green marked with dark pattern. Antennae like 2 small spikes at front of head. Dark-patterned wings held out at sides forming triangle. Horse flies (*Tabanus*) have similar habits but are larger with broader heads and all-dark, unpatterned wings.

**Habitat**  Near water in woodlands, meadows, and gardens.

**Range**  Throughout North America.

**Life Cycle**  Eggs are laid on vegetation above water; predatory larvae lie in shallow water or mud and pupate in mud at edge of water. One generation a year, but adults of different species fly during all warm months.

### Vinegar Fly *Drosophila melanogaster*

If you come upon one of these tiny, red-eyed flies in your house, take a moment to reflect on the fact that we owe much of what we know about heredity and genetics to classic experiments with this species. The "fruit fly" is very common and can easily be bred in small containers. It has a rapid life cycle so produces numerous generations in a short time, as well as a number of mutants that can be studied in the laboratory. It also has large chromosomes that are ideal for genetics research.

Identification   ¹⁄₁₆″. Very small. Dull yellowish with red eyes and dark bands on abdomen. Bristles on antennae long and plumelike. Abdomen rounded in male, pointed in female.

Habitat   Moist areas with decaying vegetation; on rotting fruit.

Range   Throughout North America.

Life Cycle   Female lays up to 2,000 eggs in decaying fruit pulp; larvae hatch in 2 days, feed on yeast, crawl out to pupate in 6 days, and emerge as adults 5 days later. Many generations a year.

### Black Flies *Simulium*

Female black flies bite fiercely and are very annoying when they swarm around a hiker's head. Even using insect repellent does not solve the problem entirely. A repellent stops black flies from landing on the skin, but they still hover an inch or so away, and this can be almost as irritating as being bitten. African herdsmen carry sections of smoldering rope during black fly season; the smoke keeps the flies at a distance.

Identification    1/16–1/8″. Stocky, hump-backed, gray or blackish flies with short, broad wings and short antennae.

Habitat    Near streams in wooded or mountainous country.

Range    North America except Florida.

Life Cycle    Females lay eggs on leaves or stones over clear running water; larvae drop into water, attach themselves to rocks, and feed on microscopic animals and plants; pupae attached to rock; emerging adults rise quickly to surface in bubble of air and fly off. One generation a year. Adults of most species fly in spring or early summer.

**House Fly** *Musca domestica*

Once these notorious spreaders of disease have mated, the males have done all that is expected of them. In about two weeks, they literally wear out and die, their flimsy wings torn and useless. But with mating, the task of the females has just begun, for they must now hunt carefully for good places to lay eggs. With all this work to do, females are more durable and live for nearly a month.

Identification ⅛–¼". Stocky and gray with 4 black lines down thorax. Abdomen gray or yellowish gray with dark line down middle and darker areas along sides. Eyes dull reddish.

Habitat In and around houses and farms, near garbage, exposed food, or manure.

Range Throughout North America.

Life Cycle Female lays eggs in clusters on garbage or manure; eggs hatch within a day; larvae mature in 5 days, pupate and emerge as adults in another 5 days. Many generations a year.

**Blow Flies** *Family Calliphoridae*

Common wherever there is decaying matter for their larvae to feed on, blow flies are easily recognized by their bright metallic colors. A few other flies are also metallic and distinguishable only by minor technical details, but if you see a bright green or blue fly that is the size and shape of a House Fly, you can be almost certain that it is a blow fly. The scavenging larvae feed on carrion and garbage; the adults are fond of sweets and regularly visit flowers to drink nectar.

Identification   ¼–⅝″. Like House Fly but metallic blue, green, or glossy black, often with reddish eyes.

Habitat   Woodlands, pastures, gardens, roadsides, and towns.

Range   Throughout North America.

Life Cycle   Clusters of eggs are laid on carrion, garbage, or other decaying matter; larvae develop rapidly and leave carrion to pupate in burrow in ground; adults emerge in 2–3 weeks. A few species are blood-sucking parasites of mammals. Many generations a year.

124

**Fireflies** *Family Lampyridae*

On evenings in early summer, fireflies can be seen blinking their lights, the males buzzing unsteadily through the air, the females answering from hiding places on the ground. But not all members of this group have light organs. Some are day-flying. Among California species, the females glow but none of the winged males do.

Identification  ¼–½". Adult males elongated, with soft, dark wing covers and usually with pale light organ at tip of abdomen; roof of thorax, which overhangs head, often marked with red or yellow. Females flightless, often resembling larvae. Larvae flattened, luminous, and obviously segmented. Females and larvae often called glow-worms.

Habitat  Open woodlands, meadows, and gardens.

Range  Throughout the United States and S. Canada.

Life Cycle  Eggs are laid on ground; larvae active, preying on snails and other small animals, hibernate, then enter pupa stage; adults emerge in summer. One generation a year.

### Eastern Box Elder Bug *Leptocoris trivittatus*

The favorite host plant of this colorful bug is the box elder, and where these trees are abundant, as they are along streams in the Midwest, spectacular swarms of females may appear in the fall looking for a place to spend the winter. They sometimes enter buildings or gather in great masses on the trunks of trees. The closely related Western Box Elder Bug (*L. rubrolineatus*), found west of the Rockies, has similar habits.

| | |
|---|---|
| Identification | ⅜–⅝″. Oval with flattened back and long "beak" and legs. Blackish with bold red markings on thorax and wings. Nymph similar but bright red and without wings. |
| Habitat | Forests, wooded stream banks, and gardens. |
| Range | North America east of Rockies. |
| Life Cycle | Eggs are laid in spring in crevices in bark of box elder, other maples, and fruit trees; nymph feeds on plant juices and transforms into adult in late summer or early fall. Male dies after mating, and female hibernates. One or 2 generations a year. |

128

**Scarlet-and-green Leafhopper** *Graphocephala coccinea*

Most of our more than 2,500 kinds of leafhoppers are small and green or drab brown. They attack only a few kinds of plants and can be crop or garden pests, but the large and colorful Scarlet-and-green Leafhopper is seldom abundant enough to cause noticeable damage to plants. It feeds on a wide variety of greenery and is most frequently seen on vegetables and ornamental shrubs such as forsythia. Like other leafhoppers, it usually attacks the leaves of plants, using its hollow "beak" like a straw to drink plant fluids.

Identification    ⅜". No other large leafhopper has bright green thorax and wings with red or orange edges and diagonal bands. Nymphs are bright yellow.

Habitat    Gardens and meadows.

Range    E. North America and S. Canada.

Life Cycle    Eggs are laid in early spring in plant tissue, and there are several generations a year. Adults overwinter on ground in leaf litter.

130

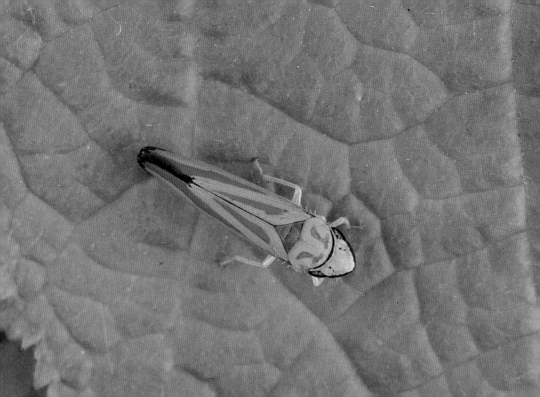

## Stink Bugs *Family Pentatomidae*

These common bugs get their name from the strong and unpleasant odor they produce when disturbed. Most species are concealingly colored. Relying both on their odor and color defenses against predators, stink bugs are often slow-moving. Although they can fly easily, they sometimes try to creep away into the foliage if you come too close. They are also called shield bugs because of their shape.

Identification ¼–¾". Flattened, shield-shaped bugs with large, triangular plate (scutellum) in middle of back. Usually green or brownish, but some species boldly patterned. Nymphs resemble adults but are smaller and lack wings.

Habitat Gardens, orchards, croplands, forests, and meadows.

Range Throughout North America.

Life Cycle Barrel-shaped eggs are laid in clusters on leaves; nymphs feed on plant juices or, in some species, on body fluids of other insects. One to 3 generations a year.

## Tiger Beetles *Cicindela*

Most beetles are slow-moving and methodical but not the tiger beetles, agile predators that can run swiftly over the ground and fly with great speed in pursuit of other insects. The larvae, too, are predators but are more stationary. They lurk at the entrance of their burrows, snatching up any insect that wanders close to their hiding place.

Identification    ⅜–⅞". Usually bright metallic green, blue, purple, or bronze, often with white spots on wing covers. Legs long. Jaws long, jagged, and sickle-shaped.

Habitat    Open sandy or dusty ground, sunny clearings, or forest paths.

Range    Throughout North America.

Life Cycle    Females deposit eggs in pits in ground; larvae hatch and dig vertical burrows; pupate in burrows and emerge as adults in following summer. One generation a year.

**Ground Beetles** *Family Carabidae*

The best way to see these common beetles is to go out at night with a flashlight; except in a desert, any dark beetle you find running around on the ground will almost certainly be a "carabid." They are very active as they search for insect prey and dart for cover if you disturb them. Darkling beetles (Family Tenebrionidae) are similar but often duller, and their jaws do not project forward; they are more common in desert areas.

Identification  $\frac{1}{12}$–$1\frac{1}{3}$". More or less flattened. Head narrower than thorax with large eyes and jaws protruding forward. Legs long and slender. Black or glossy blue, purple, or green, sometimes with red or yellow markings. Wing covers often finely grooved.

Habitat  On ground or under logs, stones, or leaves in moist areas.

Range  Throughout North America; scarce in deserts.

Life Cycle  Eggs are laid singly, often in cells, in mud or damp soil; larvae feed on other insects, pupate, and emerge as adults in following year. Adults live 2–4 years.

**Boll Weevil** *Anthonomus grandis*

The original home of the notorious Boll Weevil was Mexico or Central America, and it crossed the Rio Grande into the United States about 1890. From this early foothold, it spread into the Cotton Belt at a rate of 60 miles a year. These are weevils of many notable achievements, having become very destructive pests in cottonfields and the subject of at least one popular song. They are expert botanists, too; the adults also recognize and attack okra and hollyhock, both relatives of the cotton plant.

Identification ⅛–¼". Dark gray or brown, covered with yellowish hairs. Snout about half length of body. Wing covers grooved and pitted. Best told from numerous other weevils by its habitat and food plant.

Habitat Cotton fields.

Range S. United States.

Life Cycle Female lays eggs singly in cotton boll; larvae feed on cotton fibers and pupate in boll; adults hibernate on ground in litter. Up to 7 generations a year.

138

### European Earwig *Forficula auricularia*

When darkness comes and the hover flies, bees, and butterflies go under cover, the insect night shift takes over at flowers. This is when the European Earwig can be found most easily, munching away on pollen and petals. Its pincers are not just ornaments; they can deliver a painful pinch if you try to handle one of these nocturnal visitors.

Identification ½″. Shiny and brown or black with yellowish legs, antennae, and wings. Wing covers much shorter than abdomen. Pincers of females nearly straight; those of males curved.

Habitat Leaf litter on ground, damp places under boards and logs, and in buildings.

Range E. Canada and NE. United States; also in Pacific Northwest.

Life Cycle Up to 30 eggs are laid in burrow in spring and are guarded by female until young finish first molt. Young look like miniature editions of adults.

### Pinacate Beetles *Eleodes*

These beetles usually emerge at dusk, but you can sometimes find them lumbering around in broad daylight. If you disturb one it immediately stops, spreads its legs, and points its abdomen skyward. This is a warning well known to small rodents and other predators where pinacate beetles are common. If you continue to bother it, the beetle will release a blackish, foul-smelling fluid. Pinacate beetles are also called olive beetles because they look something like black olives, and, for obvious reasons, another name is "stink beetles."

Identification  ⅓–1½″. Stocky and black, sometimes with red stripe down center of back. Legs long and stout. Some species have fused wing covers and are flightless.

Habitat  Deserts, open grassy areas, and dry woodlands.

Range  W. North America.

Life Cycle  Females lay clusters of eggs in soil; larvae feed on roots, seeds, and other plant matter, then pupate in cell in soil. Adults may live up to 3 years and lay eggs once or twice a season.

### Eyed Click Beetles *Alaus*

All click beetles have a little hinge mechanism on the underside of the thorax that enables them to bend the body with a sudden snap. This sends them spinning into the air and enables them to right themselves when they are stranded on their backs. It also provides a handy means of recognizing a click beetle. Most species are small, dark, and hard to tell apart, but these big, "eyed" ones are easy to identify.

Identification  1–1¾". Adults elongated, gray, and mottled, with 2 eyelike black spots on thorax; rear corners of roof of thorax pointed. Larvae long and slender with thickened skin.

Habitat  Woodlands, especially near logs and dead trees.

Range  Throughout North America.

Life Cycle  Females lay eggs in soil; larvae ("wireworms") feed on plant matter or other insects and can be agricultural pests; they pupate in cell in soil or decayed wood. Adults fly all summer. One generation a year.

### Cottonwood Borer *Plectrodera scalator*

Most of North America's 1,200 kinds of long-horned beetles can be recognized easily by their very long antennae and often bold patterns. All have larvae that bore in wood. The Cottonwood Borer is a familiar species on river floodplains, where fast-growing cottonwoods and poplars are among the first trees to appear. You can usually find adults feeding on pollen at flowers in summer and early fall.

Identification    ⅞–1⅝". Elongate and glossy black with whitish or yellowish pattern formed by fine, dense hairs. Antennae much longer than body in males, slightly longer in females.

Habitat    Riverbanks with cottonwoods or poplars.

Range    E. North America west to Mississippi Valley and N. Great Plains.

Life Cycle    Eggs are laid in fall in poplar or cottonwood; larvae spend 2–3 years boring through wood near ground level, sometimes killing tree; adults emerge in midsummer.

### Harlequin Cabbage Bug *Murgantia histrionica*

This well-known crop and garden pest, one of the stink bugs, first invaded the United States from Mexico in the 1860s and has now spread over much of North America. Both adults and nymphs produce white or yellow blotches on the leaves of host plants. The bold markings vaguely resemble those of a box turtle, and the species is sometimes called the terrapin bug. Among its many other names are "calico bug" and "fire bug."

Identification   ⅜". Shield-shaped, shiny, and black, boldly marked with yellow, orange, or reddish. Nymphs similar but rounder, without wings. Eggs barrel-shaped with black and white bands.

Habitat   Gardens, orchards, and croplands.

Range   Throughout the United States and S. Canada.

Life Cycle   Adult hibernates in leaf litter, emerges in spring and glues eggs in rows on leaves of cabbage, cauliflower, mustard, and other plants; nymphs feed on plant juices. One to 4 generations a year; sometimes breeds year-round in southern states.

## Ladybird Beetles *Family Coccinellidae*

These are among the commonest and most familiar beetles in North America. Both adults and larvae prey on aphids and are useful in controlling these garden pests. In the fall, many species of "ladybugs" gather in large numbers to hibernate and sometimes enter houses. When the first warm days of spring arrive, they may appear at sunny windows, trying to escape into the outdoors.

Identification ⅛–⅜". Domed shape and color scheme—red or orange with black spots on wing covers—identify most of these beetles. Thorax often patterned black and white. Occasional individuals are black with red spots.

Habitat Gardens, meadows, and thickets; sometimes indoors in winter.

Range Throughout North America.

Life Cycle Eggs are laid in clusters near aphids. Larvae black and bristly, often spotted yellow and white. Pupae black and spotted; attached to leaf. Several generations a year.

### Spotted Asparagus Beetle *Crioceris duodecimpunctata*

Like many an agricultural pest, this beetle is not a native-born American. The first ones arrived in Maryland from Europe in 1881. The larvae cause little trouble, but adults damage asparagus leaves and spears. An even more destructive species, called simply the Asparagus Beetle (*C. asparagi*), attacks tender spears both as an adult and as a larva; it has a metallic blue head, a red thorax, and blue-black wing covers with large yellow spots and red margins.

Identification   ¼–⅜″. Adult elongate and shiny with reddish-brown head and thorax. Wing covers tan or reddish brown, each with 6 small black spots. Legs and antennae black. Larva orange with black head and legs.

Habitat   Gardens and asparagus beds.

Range   Throughout North America.

Life Cycle   First adults appear in spring; female lays eggs on asparagus; larvae feed on ripening berries, then pupate in ground. Up to 5 generations a year.

152

### June Beetles *Phyllophaga*

These sturdy, shock-resistant insects are called June beetles, May beetles, or even June bugs, because that is when the adults begin to fly clumsily about, slamming into window screens, noisily colliding with outdoor lights, ricocheting off porch walls, and clattering across the floor. This nightly bombardment is caused mainly by males, since the females of many species have small wings and cannot fly. A female must wait in the shrubbery for one of these ardent aviators to make a crash landing near her.

Identification    ¾–1⅜". Robust, shiny, uniform brown or blackish beetles. Antennae have 3 flaglike plates at tip. Body covered with fine hairs. Larvae whitish and C-shaped with brown head.

Habitat    Gardens, meadows, and open woodlands.

Range    Throughout North America.

Life Cycle    About 50–100 eggs laid in small chamber in soil; larvae feed on roots, take 2–3 years to mature, and pupate in cell close to surface; adults emerge in late spring.

### Japanese Beetle *Popillia japonica*

Every eastern gardener knows the Japanese Beetle, which reached our shores in 1916 and quickly became a major pest. The larvae feed on plant roots and can ruin a lawn or golf course; the adults feed on the leaves, flowers, and fruit of over 200 species of plants. A few decades ago many gardens produced more beetles than flowers or vegetables, but in recent years this species has been brought under control with the help of soil bacteria that kill the larvae.

Identification    ⅜–½". Adults oval, metallic green with coppery orange wing covers and white spots along each side of abdomen below wing covers. Larva, to ¾", whitish with brown head; rests in C-shaped position in soil.

Habitat    Gardens, croplands, open woodlands, and meadows.

Range    E. United States and SE. Canada.

Life Cycle    Adult lays eggs in ground in late summer; larva feeds on plant roots, hibernates deep in soil, moves to surface and pupates early in following summer. Cycle takes 2 years in North.

### Tortoise Beetles *Family Chrysomelidae*

Among the many leaf beetles of this family are a few common species that look like miniature golden turtles. When danger threatens, these beetles stop feeding and flatten themselves against the surface of a leaf so their legs and head are protected; they even withdraw their antennae out of harm's way. The larvae protect themselves by attaching debris to a long fork on the abdomen; after a while, the larvae are completely concealed by the debris.

Identification    ⅕–½″. Oval or circular and tortoise-shaped, with flattened edges that cover head and legs. Often golden or brassy, sometimes with black markings. Larvae carry debris on long fork at tip of abdomen.

Habitat    Gardens, weedy meadows, and roadsides.

Range    Throughout North America.

Life Cycle    Adults overwinter and lay eggs on leaves of host plant; larvae feed on underside of leaves, then drop to ground or attach to leaf to pupate.

### Whirligig Beetles *Gyrinus*

A group of whirligig beetles resting on the surface of the water looks like a little flotilla of watermelon seeds or coffee beans. If disturbed, they start milling around rapidly. This behavior, suggested by the word "whirligig," serves to confuse predators. If you are fast enough to catch one of these beetles by hand, you're in for a surprise: Most species give off an odor like the scent of apples. Adult whirligig beetles capture small insects that have fallen onto the surface of the water; the larvae chase down aquatic insects and mites.

Identification ⅛–¼". Oval, flattened, glossy blackish beetles with long, slender front legs and short, paddlelike hind legs. Aquatic larvae long, slender, and leggy.

Habitat On surface of lakes, ponds, and quiet pools in streams.

Range Throughout North America.

Life Cycle Eggs are laid in clusters on submerged plants; larvae are predaceous like adults; pupate in mud cells on rocks or vegetation. One generation a year.

160

You have to look closely to see that these droll little insects swim on their backs, unlike the similar water boatmen (Family Corixidae). Backswimmers spend a lot of time dangling from the surface taking in air at the tip of the abdomen. If disturbed, they paddle energetically down and out of sight. They capture aquatic insects, insects stranded on the surface, and even tadpoles by drifting up to them from below. They have a "beak" that they use to sting their prey and are not above stinging people if handled carelessly.

Identification  ¼–⅝". Elongated and boat-shaped aquatic insects with convex back (held downward) and very long, jointed, oarlike hind legs.

Habitat  Marshes, weedy ponds, and sluggish streams.

Range  Throughout North America.

Life Cycle  Eggs are cemented to submerged vegetation, and nymphs, like adults, swim on their backs and feed on aquatic insects and insects stranded on the surface. One or 2 generations a year.

162

## Water Striders *Gerris*

The slim, aquatic water striders spend most of their time skating around on the surface of quiet streams and ponds, capturing small insects that fall into the water. At egg-laying time, a male picks a safe spot on the bank and signals to passing females by creating ripples. When a female arrives, mating takes place, and the eggs are laid while the male stands guard. Most water striders are wingless, but a few species have wings and can show up on any body of water, even a mud puddle. Water striders hibernate among dry leaves near shore and return to the water on the first warm days of spring.

Identification  ⅜–1″. Flat and slender with middle and hind legs long, front legs short and adapted for grasping. Usually dark on back, whitish below. Nymphs similar but smaller.

Habitat  On surface of lakes, ponds, and pools in streams.

Range  Throughout North America.

Life Cycle  Females lay eggs in rows on stones and objects at edge of water. Nymphs mature in about 5 weeks, growing larger with each molt. Up to 3 generations a year.

### Froghoppers *Family Cercopidae*

Often called spittlebugs because of their frothy white nests, young froghoppers are much more likely to attract your attention than the adults, which feed quietly in grass or other vegetation. Both the adults and nymphs look like tiny frogs, a resemblance the adults enhance by hopping quickly away if disturbed. The nymphs are more placid and prefer to creep resignedly back into their bubble nests if you tease them out into the open.

Identification ⅛–½". Adults short, squat insects clad in gray, brown, or green; similar to some leafhoppers but stockier, without row of spines along hind legs. Nymphs pale and wingless, hide in small mass of white foam.

Habitat Fields, croplands, meadows, and thickets.

Range Throughout the United States and S. Canada.

Life Cycle Eggs are deposited on plants in fall; nymphs hatch in spring, form small masses of foam for protection, and feed on plant juices. After several molts, adults emerge. One to 3 generations a year.

## Grass Spiders *Agelenopsis*

A grass spider's flat web is not sticky, but if an insect happens to land on it, the spider can dart swiftly out of a tunnel-like hiding place and seize its prey. The tunnel is more than a hiding place; the back end is open, and if anything dangerous shows up, such as a spider-hunting wasp, the spider can slip quietly out the back door and disappear into the surrounding vegetation.

**Identification** Female ¾″, male ⅝″. Best identified by web. Usually dull brown, often striped on back. Similar to wolf spiders but have a pair of very long spinning organs at end of abdomen. Eyes 8, in 3 rows (2, then 4, then 2).

**Habitat** Meadows, lawns, and low vegetation.

**Range** Throughout North America.

**Life Cycle** Females hide eggs in fall in silken cases under bark or in vegetation, not in web. Eggs hatch during winter or in spring, and adults mature by late summer.

### American House Spider *Achaearanea tepidariorum*

The "cobweb" of this common indoor spider seems to be disorganized, but it is really a sophisticated trap for snaring insects that walk under it. The threads attached to a shelf or wall are sticky and elastic, and when an insect blunders into them, they break and hoist the victim up into the web where it can't grab hold of anything to pull itself free. One of these webs even captured a mouse, which the spider duly sampled.

Identification    Female ¼″, male ⅛–¼″. Bulbous, yellowish-brown abdomen with gray and black streaks and long banded legs make female easy to identify. Male much smaller with orange legs.

Habitat    In buildings and caves.

Range    Throughout North America.

Life Cycle    Eggs are placed in brown, hanging cocoon in web. When young hatch, they swarm about the web for a few days, then leave to start small webs of their own. Male and female often live together in the same web.

**Wolf Spiders** *Family Lycosidae*

Since she doesn't live in a web like most spiders, a typical female wolf spider must carry her eggs around with her in a ball of fine silk. She is a devoted mother, and her young ride on her back for several weeks or months after they hatch. But when the time comes for the spiderlings to fend for themselves, they had better leave in a hurry because she may forget they are her own young and pounce on them as if they were prey, a fate that probably befell their father after mating.

Identification  ⅛–1⅜″. Rather flat, nimble spiders with long, hairy legs. Usually seen running on ground. Variously colored, usually blackish or dark brown. Eyes 8, in 3 rows (4 small eyes in front, then 2 large eyes, then 2 small eyes).

Habitat  Widespread and common on the ground in all habitats.

Range  Throughout North America.

Life Cycle  Females place eggs in cocoon of silk, usually carried until eggs hatch; young then carried on back.

### **Black Widow Spider** *Latrodectus mactans*

This well-known, venomous spider gets its name "widow" from a trait that is common among spiders: Once mating has taken place, the female kills and eats the male. Although this seems cruel to us, the male is useless after mating, and the best thing he can do to insure the survival of his young is to provide a hearty meal for their expectant mother. The female hangs upside down in her web, enabling you to see her red warning signal.

Identification  Female ⅜″, male ¼″. Female has long, slender legs and spherical abdomen; glossy black with red markings (often shaped like an hourglass) on underside of abdomen. Male smaller and has more slender abdomen with red and white markings on sides.

Habitat  In sheltered places on or near the ground; in brush piles, under boards, or in cellars.

Range  Throughout S. United States.

Life Cycle  Female produces as many as 3 pear-shaped egg cases in web; young disperse after hatching.

## Black-and-yellow Argiope *Argiope aurantia*

Most spiders are timid and wear concealing colors, but a Black-and-yellow Argiope on her web behaves as though she actually wants to be seen. There she sits, in full view in the center of her orb web, on a swatch of white silk. She may even bob up and down if you come too close. Perhaps she is relying on an ingrained dislike of spiders to keep you from walking through her web and ruining the work of many hours. To deceive predators out hunting for spiders, she confuses the issue by wearing yellow and black—the colors of a yellow jacket.

Identification   Female ¾–1⅛", male ¼–⅜". Black, body marked with yellow or orange. Long legs banded with yellow or reddish.

Habitat   Tall plants in fields, meadows, and gardens.

Range   Throughout the United States and S. Canada.

Life Cycle   Female places eggs in large (1"), round sac at edge of web; eggs hatch in fall and overwinter in sac. Young leave web in spring. Male often lives in smaller web at edge of female's.

176

**Guide to Orders**

Insects and spiders are classified in groups called orders; members of an order share many basic features that help in identifying them. Usually they have the same sort of life cycle, the same kind of body, the same number and kind of wings, similar antennae, and the structures around the mouth—the mouthparts—are generally the same. There are several orders of insects, but only one of spiders. To help you understand their evolution, the orders of insects are arranged here in scientific sequence, beginning with the more primitive and ending with the more advanced.

The Insects

An insect's body has three parts: the head, thorax, and abdomen. There are two antennae on the head, and the eyes are large and usually compound—made up of many small lenses. The basic insect mouthparts include two jaws, or mandibles, that are useful for chewing, but in some orders the mouthparts are modified for feeding in other ways, and nothing resembling jaws can be seen. The thorax bears six legs, and may also bear one or two pairs of wings.

Development after hatching is gradual in some orders; the young, called nymphs or naiads, are similar to adults, but acquire wings only at maturity. In other orders the life cycle has four distinct stages: the egg; the young, called a larva, which does not resemble an adult; an inert, resting stage called a pupa; and finally the adult.

**Mayflies**

Adult mayflies (order Ephemeroptera) are slender insects with two or four more-or-less clear wings with many veins; the wings are held like sails above the back. There are long threads at the tip of the abdomen. Adults have weak mouthparts, and most are short-lived and do not eat. Development is gradual: The aquatic naiads live in streams or lakes and often form great swarms when they emerge as adults. (page 76)

**Dragonflies and Damselflies**

Aerial hunters, dragonflies and damselflies (order Odonata) have slender bodies and four many-veined wings. Dragonflies hold the wings out to the sides, while damselflies fold them over the abdomen or keep them somewhat fanned. These insects have large compound eyes and keen vision. The antennae are very short. Their long legs, unsuitable for walking, are meant to hold prey. The aquatic naiads feed on small insects, tadpoles, and fish; they leave the water to transform into adults. (pages 70–74)

**Mantids**

The mantids (order Mantodea) have powerful front legs that are armed with spines and used to grasp prey. They

179

are green or brown with slender bodies and small, triangular heads with very short antennae. They have powerful jaws and two pairs of wings; the first pair of wings is leathery. Mantids develop gradually. (page 30)

Termites

Living in teeming colonies underground or in wood, termites (order Isoptera) are small insects with soft bodies and chewing jaws. Development is gradual, and it is not always easy to tell if a large nymph is fully grown or not. The sterile workers are wingless, but fertile males and females have four very delicate wings that are shed after a brief mating flight. (page 94)

Earwigs

The earwigs (order Dermaptera) are a small group of flattened insects with chewing mouthparts and a pair of curved pincers at the end of the abdomen. The four membranous wings fold up onto the back when not in use, but many earwigs are wingless. Females tend their eggs in underground tunnels, and development is gradual. (page 140)

Stoneflies

Adult stoneflies (order Plecoptera) have long, threadlike antennae and two pairs of membranous wings that fold over the abdomen and often seem to wrap around it. Most stoneflies have long antennalike cerci at the end of

180

the abdomen. The mouthparts are small, and many adults are short-lived and do not eat. The aquatic naiads live in oxygen-rich streams. (page 80)

Grasshoppers and Crickets

Members of this group (order Orthoptera) are easily recognized by their hind legs, which are long, stout, and used for jumping. Grasshoppers, crickets, katydids, and their allies have strong jaws for chewing foliage, and most species have four wings, with the fore wings leathery and the hind wings more-or-less clear. The head is large, and the antennae may be very long and slender or short and peglike. The females of many species have a long, bladelike ovipositor for inserting eggs into plant tissue or the ground. The young are like adults but smaller and without wings. (pages 18–24, 28)

True Bugs

The true bugs (order Hemiptera) have two pairs of wings; the first set of wings are leathery at the base, and fold flat over the back on either side of a triangular plate called the scutellum. These insects have a "beak" that juts out at the front of the head; they suck plant juices and sometimes feed on other insects. The nymphs may have small wing buds when they are young, but are able to fly only at maturity; they usually feed in the same way as the adults. (pages 128, 132, 148, 162–164)

181

**Cicadas, Leafhoppers, and Kin**

Cicadas, leafhoppers, and their relatives (order Homoptera) are similar to the true bugs, but their wings are usually clear and veined, and the "beak" arises far back under the head, where it can be hard to see. All are plant-feeders and some are very destructive agricultural pests. The young usually look like wingless adults. (pages 26, 62–64, 130, 166)

**Lacewings**

Lacewings (order Neuroptera) are weak-flying and mainly nocturnal. They have threadlike antennae and large but delicate wings with numerous veins. Most are green, and their eyes are brilliant gold. Both adults and the active larvae have chewing mouthparts and prey on aphids. The pupa has a silken cocoon. (page 68)

**Beetles**

Although beetles (order Coleoptera) come in many shapes, they are easily recognized by their two hardened front wings, which serve as wing covers to protect the more delicate and membranous hind wings that do the actual flying. The antennae are more variable than in any other order, ranging from very long and slender to short and peglike. Both adults and larvae have chewing mouthparts. The larvae feed in many habitats and are highly variable; some are legless grubs, while others are agile predators with six legs and sharp jaws. The pupa is

182

attached to a leaf or lives in a cell in the soil, in wood, or under water; there is no cocoon. (pages 126, 134–138, 142–146, 150–160)

Scorpionflies

The scorpionflies (order Mecoptera) are odd-looking insects with long legs, a long snout with chewing mouthparts, long and slender antennae, and usually with four pairs of dark or mottled, membranous wings. They are common in damp places. The larvae resemble caterpillars and live in moist soil. (page 82)

True Flies

The true flies (order Diptera) have only one pair of functional wings; the second pair is reduced to two small structures shaped like dumbbells that are thought to aid in balancing. These insects have lapping or piercing mouthparts, and short antennae that are feathery in a few kinds of flies. They are termed "true" flies to distinguish them from the many other insects that are called flies but do not have a single pair of wings, or have other kinds of mouthparts. Most fly larvae are legless and headless; they live in damp places or in water. The larvae usually crawl to a drier place to transform into a pupa. (pages 78, 84–88, 106, 114–124)

Caddisflies   The caddisflies (order Trichoptera) are dull-colored,

183

nocturnal, mothlike insects with four opaque wings, long, threadlike antennae, and chewing mouthparts. They are usually found near water because the larvae are aquatic, building protective cases of sand or bits of debris, and pupating in their underwater cases. (page 66)

**Butterflies and Moths**

Familiar insects, butterflies and moths (order Lepidoptera) have large wings covered with scales. Their mouthparts consist of a coiled tube for sipping liquids like nectar and the sap of plants. Butterflies are usually brightly colored, and have slender, knobbed antennae; they fly during the day. In contrast, moths are mainly nocturnal, and usually have feathery antennae; day-flying moths are often brightly colored, but the nocturnal ones tend to be clad in concealing browns and grays. The larvae are called caterpillars; those of many moths weave a silken cocoon when they change to a pupa. A butterfly pupa does not have a cocoon and is called a chrysalis. (pages 32–60)

**Wasps, Bees, and Ants**

The wasps, bees, ants, and kin (order Hymenoptera) are mostly social species that live in colonies, tending the eggs, larvae, and pupae of a fertile queen. They have two pairs of wings—either clear or dark—and strong, chewing jaws. Except in horntails and their relatives,

184

the abdomen is set off from the thorax by a narrow "waist." In those members that can sting, the abdomen may be boldly colored as a warning. The antennae are usually not very long, but they are very active; those of ants have a joint in the middle and are conspicuously "elbowed." Worker ants are wingless, but fertile males and females fly during a short mating period after which the males die. Some members of this order are not social but solitary, and the females make nests by themselves. Others, such as the ichneumons, are parasites that lay their eggs in the bodies of other insects. All members of this order go through a four-stage life cycle. (pages 90–92, 96–104, 108–112)

The Spiders  Actually, spiders are not insects but arachnids, a group that also includes scorpions, ticks, and mites. The spider's body has an upper section, the cephalothorax, and a lower section, the abdomen. There are eight legs, up to eight eyes that are never compound like those of insects, and two poison fangs. All spiders produce silk to form a sac for the eggs and, in most, to make a web for trapping prey or to line a burrow in the ground. Young spiders resemble adults. (pages 168–176)

**Abdomen**
The hindmost part of an insect's body.

**Antenna**
One of a pair of sensory organs on the head of an insect, often long and slender; also called "feelers" (pl. antennae).

**Cercus**
One of two antennalike or jawlike organs at the end of an insect's abdomen (pl. cerci).

**Chrysalis**
The pupa of a butterfly.

**Cocoon**
The silken covering that protects many insect pupae.

**Generation**
A single cycle from egg to adult. Many insects have several such cycles, sometimes called broods, during a year.

**Larva**
A young, often wormlike, insect that looks very different from the adult and that passes through a pupa stage before becoming an adult (pl. larvae).

### Naiad
A young insect that lives under water, has legs and often gills, and does not have a pupa stage.

### Nymph
A young insect that resembles an adult but lacks wings, lives on land, and does not have a pupa stage.

### Ovipositor
A spinelike or bladelike structure at the tip of the abdomen of a female insect, used to deposit eggs.

### Pupa
An inactive, resting stage during which the insect larva changes into an adult (pl. pupae).

### Pupate
To change from a larva to a pupa.

### Scutellum
A triangular plate located on the back between the bases of the wings of some insects, especially the true bugs (order Hemiptera).

### Thorax
The middle portion of an insect's body, bearing the legs and wings.

# Index

Numbers in italics refer to insects and spiders mentioned as similar species.

**The Audubon Society**

The National Audubon Society is among the oldest and largest private conservation organizations in the world. With over 560,000 members and more than 500 local chapters across the country, the Society works in behalf of our natural heritage through environmental education and conservation action. It protects wildlife in more than seventy sanctuaries from coast to coast. It also operates outdoor education centers and ecology workshops and publishes the prizewinning AUDUBON magazine, AMERICAN BIRDS magazine, newsletters, films, and other educational materials. For further information regarding membership in the Society, write to the National Audubon Society, 950 Third Avenue, New York, New York 10022.